First Life on Earth
Scientific Adam & Eve

By Walter Parks

Edited by
Laura J. Silverman

Cover is a Rubens Painting with Insert

Copyright 2013
UnKnownTruths
Publishing Company

UnKnownTruths
Publishing Company
8815 Conroy Windermere Rd. Ste 190
Orlando, FL 32835
UnknownTruths.com
info@UnknownTruths.com

Contents

Introduction

"Life is a moderately good play with a badly written third act."

Truman Capote

From the earliest days as thinking humans we have had many questions about life. We have had fewer acceptable answers. This is beginning to change as we are learning more and more about life and how it developed.

I grew up in the Bible Belt of Mississippi. We were taught that God created all life and then He created Adam and Eve in the Garden of Eden. But then I read about ancient fossils of human-like creatures and started learning about evolution.

So what is the truth? Am I to continue to believe the literal words of the Bible or the findings of archeology and the related scientific evidence?

I had to find the truth so I started my search. Let's explore what I learned.

As we walk about Earth and look around we learn that there are about ten million species of animals and several hundred thousand species of plants, but all the lives that we see are latecomers on planet Earth.

And even the fossils of old that we find, when added to the life forms of today, make up only a small fraction of species that have lived on planet Earth.

Key Life on Earth

I had thought Man was God's most important creation, and perhaps we are. But we are not the key life on earth. Earth would get along very well, perhaps even better, without us.

The necessary life on earth is the greater diversity of microorganisms: bacteria, protozoan, and algae. They make up by far the most life on Earth.

When you consider the ecological circuitry of Earth, i.e. the ways in which materials like carbon, sulfur, phosphorous, and nitrogen get cycled in ways that make them available for all biology, the organisms that do most of the real work are bacteria.

Bacteria are essential for our ecology; we are optional.

The big difference between a bacterium and us is that our bodies consist of trillions of cells that function in a coordinated manner, but bacteria are single cells; although, they're not really independent. Bacteria actually live in associations of great numbers. They are not lone operators. They work in these very highly coordinated communities of organisms that help each other grow and prosper.

To understand the first life on earth and the first man on earth, we need to explore what life really is, how it got started, and how we came to be.

Chapter 1
What is Life

"Life is an opportunity." **Mother Teresa**
"Life is DNA." **Walter Parks**

Life distinguishes objects that have self-sustaining biological processes from inanimate objects. Life is the condition which distinguishes active organisms from inorganic matter.

Living organisms on earth have carbon and water-based cellular forms with complex organizations that contain heritable genetic information.

Living organisms undergo metabolism, possess a capacity to grow, respond to stimuli, reproduce, and through natural selection, adapt to their environment in successive generations.

More complex living organisms can communicate through various means.

Earth's life forms vary in complexity from one-cell creatures such as amoeba to the most complex, which is man.

The cell is the basic unit of a living organism.

The cell is the simplest living unit of matter.

From the unicellular bacteria to multi-cellular animals, the cell is one of the basic organizational principles of biology. There are basically two types of cells: prokaryotes and eukaryotes. Prokaryote cells have no nucleus and are the basic life form for some bacteria and algae.

Prokaryote Cells

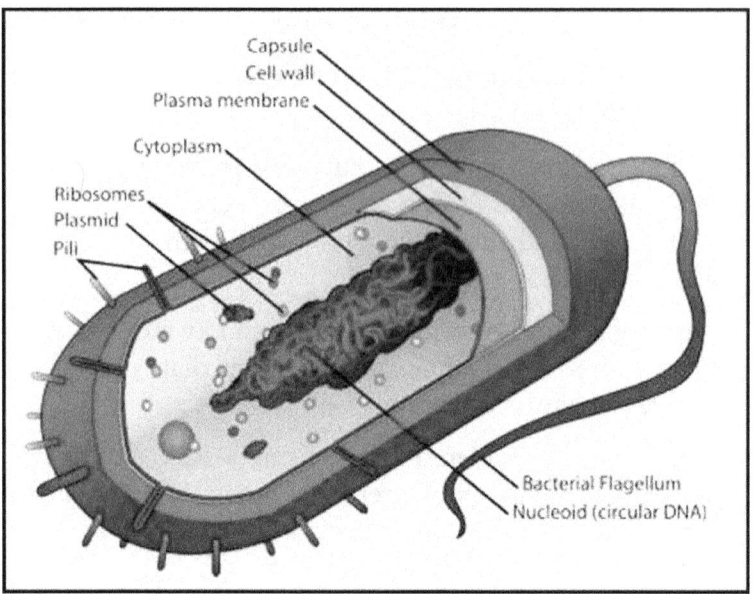

Prokaryotic Cell Structure

Prokaryotes are single-celled organisms that are the earliest and most primitive forms of life that are still on earth. Prokaryotes include bacteria and related microorganisms. Prokaryotes are able to live and thrive in various types of environments, including extreme habitats such as hydrothermal vents, hot springs, swamps, wetlands, and the guts of animals.

Prokaryotic cells are not as complex as eukaryotic cells. They have no true nuclei as the DNA is not contained within a membrane or separated from the rest of the cell but are instead coiled up in a region of the cytoplasm called the nucleoid. Using bacteria as our sample prokaryote, the following structures can be found in bacterial cells:

Capsule - Found in some bacterial cells, this additional outer covering protects the cell when it is engulfed by other organisms, assists in retaining moisture, and helps the cell adhere to surfaces and nutrients.

Cell Wall - Outer covering of most cells that protects the bacterial cell and gives it shape.

Cytoplasm - A gel-like substance composed mainly of water that also contains enzymes, salts, cell components, and various organic molecules.

Cell Membrane or Plasma Membrane - Surrounds the cell's cytoplasm and regulates the flow of substances in and out of the cell.

Pili - Hair-like structures on the surface of the cell that attach to other bacterial cells. Shorter Pili called fimbriae help bacteria attach to surfaces.

Flagella - Long, whip-like protrusions that aid in cellular locomotion.

Ribosomes - Cell structures responsible for protein production.

Plasmids - Gene carrying, circular DNA structures that are not involved in reproduction.

Nucleoid Region - Area of the cytoplasm that contains the single bacterial DNA molecule.

Most prokaryotes reproduce through a process called binary fission. During binary fission, the single DNA molecule replicates, and the original cell is divided into two identical daughter cells.

Eukaryote Cells

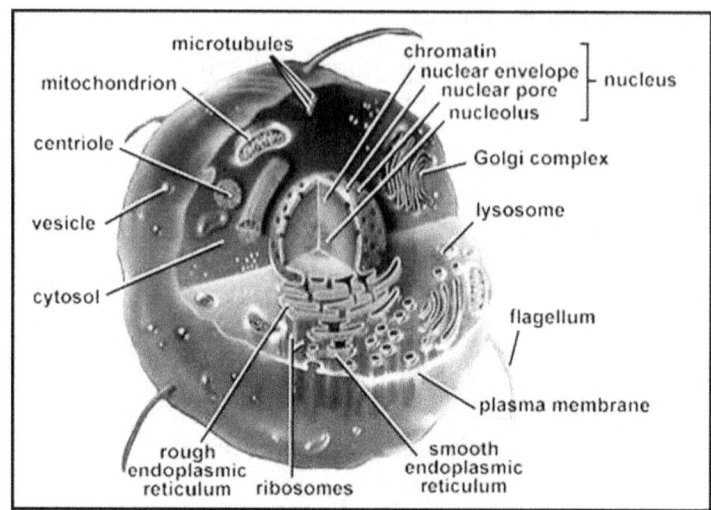

**A model of a eukaryotic cell
(Picture taken from online Biology Book)**

A eukaryotic cell has a nucleus, which is separated from the rest of the cell by a membrane. The nucleus contains chromosomes, which are the carriers of the genetic material.

There are internal membrane-enclosed compartments within eukaryotic cells called organelles, e.g. centrioles, lysosomes, Golgi complexes, and mitochondria, among others as shown above. Each is specialized for a particular biological process. The mitochondria are found in all eukaryotes and are specialized for energy production.

The area of the cell outside the nucleus and the organelles is called the cytoplasm.

Membranes are complex structures, and they are effective barriers to the environment. They also regulate the flow of food, energy, and information in and out of the cell.

There is a theory that mitochondria are prokaryotes living within the eukaryotic cells.

The eukaryote cells have a nucleus and are the basic life form of the higher order of creatures.

All cells have the same basic structure. They have an outer covering called a plasma membrane. The plasma membrane holds the cell together and permits the passage of substances in and out of the cell.

The interior of cell is called the cytoplasm. Embedded within the cytoplasm of eukaryotes are the cellular organelles, which perform the functions of the cells.

In organisms with more than one cell, a collection of cells work together to perform similar functions. This collection is called a tissue. Tissues that perform coordinated functions form organs. Organs that work together to perform general processes form body systems.

There are many types of cells. Multi-cellular organisms contain a vast array of highly specialized cells. Plants, for example, contain root cells, leaf cells, and stem cells. Examples of human cells include skin cells, nerve cells, and sex cells. Each kind of cell is structured to perform a highly specialized function.

DNA Structure of Chromosomes and Telomeres

All life on planet earth is designed and operated by its DNA, which is a very large molecule in the nucleus of each cell. The DNA caries codes that perform 3 functions:

1) Build the proteins that structure our bodies
2) Operate switches that turn the protein-making genes off and on
3) Operate the "command center" that activates the switches

We can better understand this biological nature of life by understanding the DNA structure of our chromosomes and telomeres.

DNA is a double-stranded helix made up of base pairs. It is frequently depicted as follows:

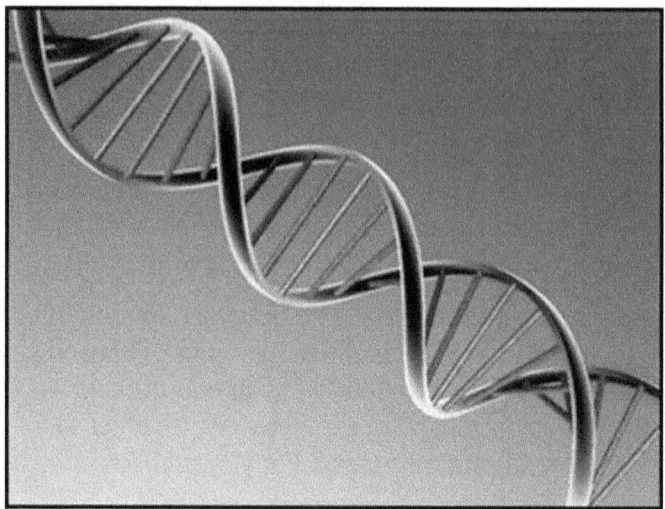

This double helical structure was discovered by James Watson and Francis Crick in 1953. Their discovery opened the door to a new and very detailed window of knowledge about the structure of life.

It's interesting to note their early statement: "This structure has novel features which are of considerable biological interest."

This is now considered to be one of science's most famous understatements. The DNA-helix is the molecule that carries genetic information from one generation to the other.

Nine years later, in 1962, they shared the Nobel Prize in Physiology or Medicine with Maurice Wilkins, for solving one of the most important of all biological riddles.

Even now, more than a half century later important new implications of this contribution to science are still coming to light.

Yes, I would say it is of "considerable biological interest!"

Let's take a closer look to better understand the nature of the base pairs.

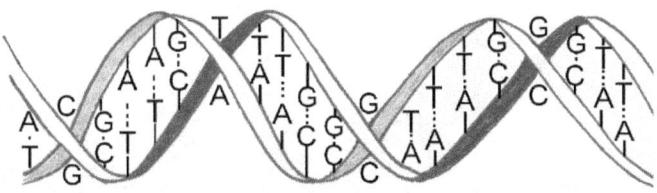

The base pairs, i.e. the steps in the DNA ladder, are made of 4 substances: adenine on one strand always pairs with thymine on the other strand. These are called A-T pairs, regardless of which strand has the "A" and which has the "T."

Similarly, cytosine on one strand always pairs with guanine on the other strand, creating G-C pairs.

Scientists often represent DNA strands with a string of letters like this:

ATATTTGAAAGCTGTGTCTGTAAACTGATGGCTAA CAAAACTAG.

This string of letters represents only one strand, or one-half of the DNA molecule. There is no need to write down the other strand because as I just described above, a "G" in one strand means there is automatically a "C" in the other strand, just as a "C" in one strand implies that the other contains a "G."

A group of these sequences make up a gene as illustrated.

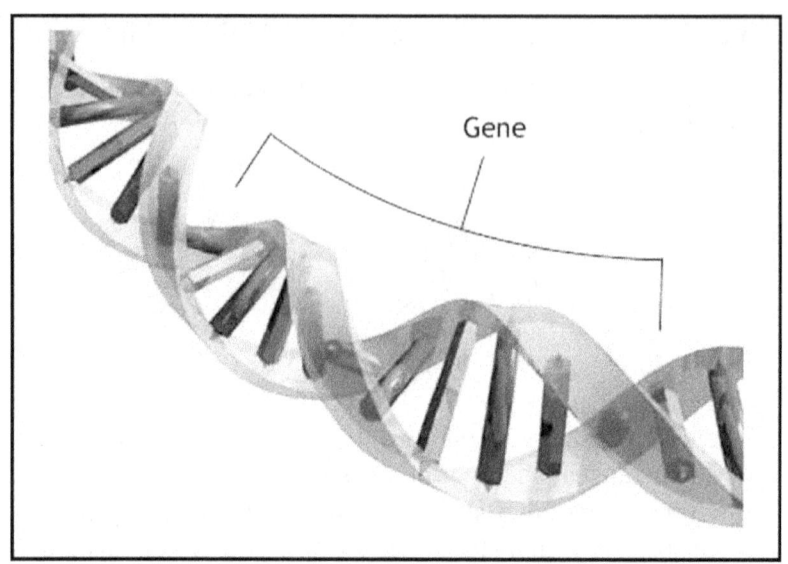

Gene

The average gene is about 27,000 base pairs long. The longest discovered gene is a muscle-production gene; it is 3.2 billion base pairs long.

At the end of each chromosome are telomeres, which are non-coding, repeating genes with the base pair sequence of TTAGGG. The telomere can reach a length of 15,000 base pairs.

Telomeres Cap Chromosomes for Protection

Telomeres function by preventing chromosomes from losing base pair sequences at their ends. They also stop chromosomes from fusing to each other.

The telomeres have been likened to the grommets at the ends of shoelaces that prevent them from fraying; however, each time a cell divides to replace itself, some of the telomere, usually about 25-200 base pairs per division, is lost.

When the telomere becomes too short, the chromosome reaches a critical length and can no longer replicate. This means that its cell becomes too old, and it dies without replacing itself.

In a simplified description of the DNA operation, the sequence of base pairs that make up a gene split in the center, exposing the ATCG bases. These bases attract materials from within the cell to replace the other side of the bases, e.g. TAGC in a structure called RNA. This becomes a protein that the body needs for its structure and growth.

The protein is made from 20 amino acids that the DNA finds within the cell. The protein may be composed of a small number of amino acids or very large numbers of the 20 basic amino acids.

The genes that operate as switches and the genes that send command messages to the switches have only recently been found and are still not fully understood. Only recently these genes, since they did not produce proteins, were thought to be junk DNA. I will skip attempting to describe them further, less I make this book more complicated than its purpose warrants.

So biological life is a cellular organism that feeds itself, undergoes metabolism, possess a capacity to grow, responds to stimuli, reproduces and passes on heritable genetic information, and through natural selection, adapts to its environment in successive generations. More complex living organisms can communicate through various means.

You can learn more details about this structure in my book: **Aging is Preventable.**

AGING IS PREVENTABLE

SEEKING ETERNITY WITH HORMONES & TELOMERES

WALTER PARKS

I wrote this book to record what I learned about how the aging process works. I, like Ponce de Leon was searching for eternal youth.

I was very surprised to learn that aging can be prevented, but that's the other story. Let's get back to **First Life on Earth**.

We now know that all life on earth evolved from very primitive DNA, as we will discuss later.

Life is the DNA that mutates, and by so doing, creates all the various species. The life of a specific species is the species DNA.

Each of us humans are the continuing and evolving life from the DNA seed planted or created on earth about 4 billion years ago.

So we see that DNA is the basic answer to our question: What is life?

Life is DNA. Yes, that's the basic answer, but we learn it's not the final answer as we explore the next question: How did life begin?

Chapter 2
How did Life Begin

"I don't know the question, but sex is definitely the answer." **Woody Allen**

Okay, so we know that all life was created or evolved from a primitive DNA seed. We can trace all life here on earth back to that first seed. But where did the first seed come from?

And how did that first seed lead to the great variety of creations now living on earth?

How did that first seed lead to the creation or evolution of man?

The answer? It's complicated. Let's begin by reviewing how non-life first became life.

Crystalline Growth

Let's start by considering crystals. They grow via the laws of physics and chemistry.

Snowflakes and ice crystals form as the temperature drops below that of liquid water; they crystallize.

They will continue to grow as long as liquid water is available and the temperature remains below the freezing point. They can also grow through sublimation, from water vapor in the atmosphere.

In a similar manner crystals of other materials form and grow.

There are many materials that form growing crystals.

I became so interested in crystals that I wrote a book about them: **Crystal Healing, Scientific Evidence.**

CRYSTAL HEALING
SCIENTIFIC EVIDENCE

WALTER PARKS

I even feature them in my foyer.

Crystals can grow via various processes including evaporation, a change in the pH (acidity) of the solution from which they grow, temperature changes, or any other changes in the solution from which they grow.

When crystals form slowly, each ion or molecule finds its correct place in the crystal lattice, and this creates an almost perfect crystal. When the precipitate is very rapid, the crystal forms and is often impure.

Okay, so we see how non-life grows. Now let's see how first life may have developed from such crystalline-like processes.

Creation of Amino Acids

In 1953, Miller and Urey speculated about what conditions may have existed on primordial Earth.

Stanley L. Miller and Harold C. Urey

They combined ammonia, hydrogen, methane, and water vapor and inserted electrical sparks to simulate the energy provided by lightning.

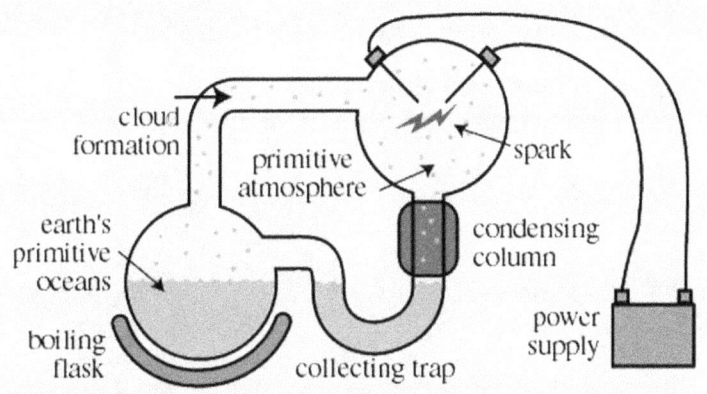

This experiment resulted in the formulation of new molecules that they then identified as the molecules of eleven of the standard twenty amino acids that make up proteins of today's life. It should be noted that a 21st amino acid of this type has recently been found, and a 22nd may soon be found.

It is also interesting to note that many of the compounds made in the Miller/Urey experiment are known to exist in outer space. Indications of this are from a meteorite that fell over Murchison, Australia on September 28, 1969. Analysis of the meteorite showed that it was rich with amino acids.

Over 90 amino acids have been identified by researchers to date. Twenty or more of these amino acids are found on Earth.

The early Earth is believed to be similar to many of the asteroids and comets still roaming the galaxy.

If amino acids are able to survive in outer space under extreme conditions, then this might suggest that amino acids were present when the Earth was formed. More importantly, the Murchison meteorite has demonstrated that the Earth may have acquired some of its amino acids and other organic compounds by impacts from space.

The typical amino acid structure is as follows.

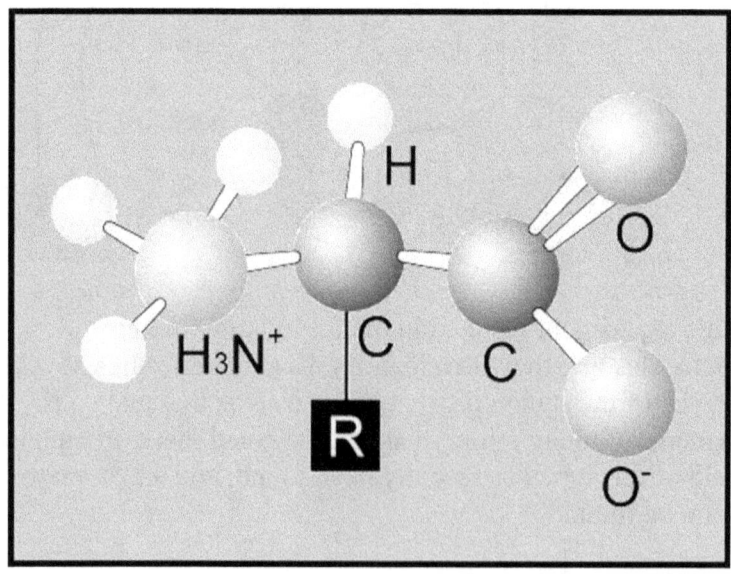

All amino acids contain three parts: an amino group H_3N+, a carboxyl group (COO-), and an R group. The R group varies among the different amino acids.

Creation of Other Life Components

The Miller/Urey experiment also produced a large amount of the nucleotide base, adenine. Adenine is of tremendous biological significance as an organic compound because it is one of the four bases in RNA and DNA.

Adenine is also a component of adenosine triphosphate, or ATP, which is a major energy releasing molecule in cells.

The creation of certain components of life is fairly easy. Simple sugars, molecules called bases, which are at the heart of DNA, and molecules called amino acids, which are at the heart of proteins, have been seen to readily form.

It's also fairly easy to make some of the fatty substances that make the coverings of cells.

Making all of these building blocks individually seems to be pretty plausible.

The hard part is how to get them to work together. How did it go from some warm pond on the primordial Earth that has amino acids, sugars, fatty acids, and more just floating around in the environment, to something in which nucleic acids are actually directing proteins to make the membranes of the cell?

I will try to describe how life began, giving my best answers, but you must recognize that the initial emergence of life was most likely an event that occurred 3.5 billion to 4 billion years ago. Even the rocks of that era have mostly vanished, so we will have to rely on some speculation to fill in the gaps between actual evidence.

Life most likely started in the simplest possible way, as a cycle, i.e. a natural chemical reaction that repeated itself, spinning off byproducts. Some of these byproducts survived to develop and maintain the cycle.

But where and how did this cycle start?

Dr. Wächtershaüser favors some mineral surface-like iron pyrites, also known as fool's gold. The iron pyrites are natural catalysts that could have assembled chemicals like carbon monoxide into biological building blocks.

Dr. Gunter Wächtershaüser Iron–sulfur world theory was proposed in a series of articles between 1988 and 1992. Dr. Wächtershaüser is a Munich patent lawyer with a degree in chemistry.

His theory proposes that early life may have formed on the surface of iron sulfide minerals. He believes that the earliest form of life originated in a volcanic hydrothermal flow at high pressure and high (100 C) temperature. It had a composite structure of a mineral base with iron and nickel and probably included cobalt, manganese, tungsten and zinc.

This beginning process is not unlike the crystal growth described above for non-life.

At some stage, the little cycle acquired a cover of protective chemicals to separate its own reactions from the general environment. When the cover eventually enveloped the cycle and broke free of the mineral surface, the first cell was born.

Recently researchers from Massachusetts General Hospital reported that montmorillonite clay, formed from weathered volcanic ash and familiar to many households as cat litter, has a property of possible relevance to the origin of life. It makes droplets of fat molecules that rearrange themselves into small bubbles, similar to the membranes that make up the walls of living cells.

Often clay particles are incorporated into the bubbles, and they could contain attached pre-RNA and RNA molecules. The researchers concluded that such mineral particles may have greatly facilitated the emergence of the first cells.

In a second experiment, the researchers found that they could make their proto-cells divide by forcing them through fine holes in a filter. A natural counterpart to this process, they suggest, would be water currents forcing bubbles through rock pores. I will later discuss the importance of how cells know when to divide.

The First Life

All of this tends to support the Miller and Urey conclusion that life's first organisms on Earth likely arose in an environment similar to the one they constructed in their lab. This environment, rich in organic compounds, is now widely described as the primordial soup.

This hypothesis is further extended to the claim that, within this soup, single-celled organisms evolved, and as the number of organisms increased, the organic compounds were depleted.

Necessarily, in this competitive environment, those organisms that were able to biosynthesize their own nutrients from elements had a great advantage over those that could not.

Today, the vast majorities of organic compounds derive from biological organisms that break down and replenish the resources for sustaining other organisms, and rather than emerging from an electrified primordial soup, amino acids now primarily emerge from biosynthetic enzymatic creations and the remains of dead organisms.

The Miller and Urey conclusions may well be how life got started at the cellular level, with simple, self-replicating molecules derived from the created amino acids. Or the amino acids required for life may have come from outer space.

I guess it really doesn't matter. The key point is that either way, the initial steps towards life appears to have been much easier than earlier realized.

The RNA World

The next steps of how these created amino acids came together to form combinations and then RNA, which developed into DNA, requires a very detailed technical discussion. I have omitted the technical discussion of these steps because it would take us down a tedious and detailed discussion of organic chemistry that would likely detract from understanding the fundamentals of how life began on Earth.

But I do need to note that biologists have long supposed that RNA was the key component in early-cell development. It could act as an enzyme, a catalyst of chemical activities, and it could also store genetic information. Most of the early RNA functions, including information storage, were later passed to DNA, which is a more stable molecule.

Some believe that this step-by-step creation from crystals to DNA is sufficient to describe how life began in that after a sufficiently long period of Darwinian evolution the humble non-life replicator cell eventually transformed into an entity complex enough that it became indisputably living.

These early biological systems can be distinguished from chemical systems because they contain components that have many potential alternative compositions, and that through trial and error, evolution can adapt them and create a molecular memory or genotype, which becomes shaped by experience, by Darwinian selection, and are maintained by self-reproduction.

Suffice it to say that Miller and Urey pretty much proved that amino acids, the building blocks of life, could almost spontaneously form in the environment believed to have existed in the early days of earth.

I accept this theory for single-cell simple life.

I can also readily accept the facts of cellular osmosis, whereby water and nutrients can be absorbed from a soup outside the cell, and that waste products can osmoses out of the cell. I therefore can see how cells could receive sufficient nutrients to grow.

You can learn more on this life-forming process in my book: **Life and the Universe**.

I wrote this book with my lifelong friend from the first grade. He and I have debated each new scientific finding as they have occurred. We had primarily discussed how the new findings affected our boyhood Christian teachings.

We decided to research all that we know about life and the universe and write it in a book. We though it would be of interest to the baby boomers and the young who are just beginning search for understanding life and the universe.

Chapter 3
Informational Molecules and Functionality

"Molecular information control is the key defining characteristic of life." **Walter Parks**

It's hard to understand how cells know when they have become large enough that they need to split and form a daughter cell. Where do they get this instruction?

Simple evolution theory does not seem to provide the answer. I began to think that I had found a chink in the armor of evolution!

But as we learn below, that was not the case.

The cells do need information to tell them how to proceed to life and in life.

So this raises a more basic question about creation or evolution: How did the first informational molecules come into existence, i.e. the molecules needed to direct the generation of efficient self-replication machinery.

The purely mechanical forcing of cells to divide as previously described may have initiated the creation of cell division instructors, such as the special proteins called cyclins as described below.

However more complex life did not really begin until there was a production of macromolecules that served as primitive stores of genetic information, i.e. until genomes that contained the informational molecules were created.

Life is more than just complex chemistry; it requires unique informational management properties that are crucial factors in escalating non-life to life.

The manner in which information flows through and between cells and sub-cellular structures is quite unlike anything else observed in chemical-only nature.

Functionality

This biological information flow gives each cell and the organism of which it is a part the functionality needed for life.

The information content of DNA is only a small part of the story. DNA is not a blueprint for an organism; DNA is mostly a passive repository for transcription of stored data into RNA, some of which goes on to be translated into proteins.

The biologically relevant information stored in DNA therefore has very little to do with its specific chemical nature. It is the functionality of the DNA expressed RNA and proteins that is biologically important.

Functionality, however, is not a local property of a molecule.

It is therefore not possible to determine what will be functional in a cell based on local structure and sequence information alone. It is thus clear that functionality, the most important feature of biological information, is decisively non-local. Functionality is subject to informational control and feedback; it changes with time in a manner that is both a function of the current state and the history of the organism.

Biological systems information is a property of the system. It is distinctive because it is the information that determines the current state, its dynamics of change, and the future state of the system.

Information control is the key defining characteristic of life.

But how did information molecules first come into being and gain control of living organisms?

Early systems without information control existed in the past, but they were unlikely to evolve over geological timescales without acquiring informational protocols. Therefore, life forms with informational control may be the only systems that evolved in the long run, and are thus the only remaining products of the processes that led to complex life.

Some very simple life that did not gain the informational molecules did survive, and some still live today. But that is not the kind of life I am interested in for this book.

Thus the onset of Darwinian evolution in a chemical system was probably not the critical step in the emergence of life. Instead, the emergence of life was probably marked by a transition to information processing capabilities.

Of major interest is the determination of how information control emerged in the pre-RNA and then RNA world, setting from just chemical kinetics to life forms with primitive information control mechanisms.

Let's look back at the simple one-cell animal and consider how it gets its food.

It must be in a nutrient-laden environment, and the chemical pressures (called partial pressures) of the water and nutrients outside its body must be greater than the pressures inside. The water and nutrients then osmoses through the cell wall and into the cell so that it can grow.

The cell does not need informational molecules to allow this to happen.

But how does the cell know when to stop growing and to split into daughter cells?

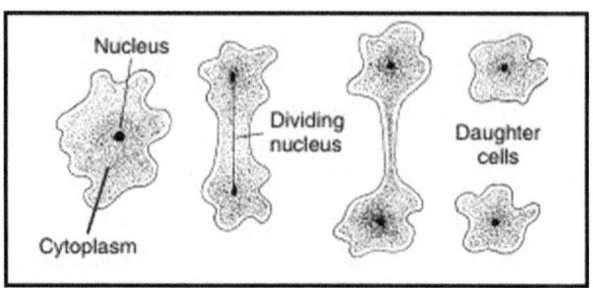

All cells follow a routine. Cells know (because of chemical pressure differentials between inside and outside) when they have to take in food and when to grow, and in the same way, they know when it is time for them to divide. This routine is called the cell cycle.

The life of a cell begins when a parent cell divides by mitosis. This process is very well-controlled by the cell. When they are dividing, cells produce special proteins called cyclins. These proteins are produced only during a particular time of the cell cycle.

If the cell grows too large there will not be enough interface between the inside of the cell and its outside (called wetted area) to provide enough cell wall surface for sufficient nutrients to be absorbed, and the cell will starve and die.

In the Darwinian evolution process over a long period of time some cells developed the ability to sense this situation, and an informational molecule was created in their DNA that tells it to divide into daughter cells. This sensing led to the creation of the molecule that produces the messenger protein cyclins as mentioned above.

Such a process of creation for various types of informational molecules in the DNA continued over time to create receptors within the cells for sensing the evolving needs for life.

This process has resulted in creating receptor sensors for the very large number of enzymes and hormones that are the informational molecules that control our bodies today.

I will not repeat the details here, but you can learn more about these informational carriers in my book: **Hormones Working For You.**

I wrote this book to record what I had learned about our hormone system.

The hormones serve as messengers to receptors in all parts of our body. Hormones control our moment to moment and all lifetime activities to, first insure survival of our species and then to allow us to continue to live long enough to raise our children to maturity.

Hormones decline rapidly after that. But we have learned that we can do something about hormone decline to increase our healthy longevity. These actions are described in my previously mentioned book: **Aging is Preventable**

Chapter 4
Steps Towards Evolution

"It is not the strongest or the most intelligent who will survive but those who can best manage change." **Charles Darwin**

So what do we conclude about how life began?

Looking at today's life forms and tracing the creation or evolution back, it's pretty obvious that all the organisms living today, even the simplest ones, are far removed from the initial life forms of about four billion years.

Those initial life forms would have been much, much simpler than anything living today. They survived and evolved because they must have had that fundamental property of being able to grow and reproduce and be subject to Darwinian evolution.

The earliest things that actually fit that definition were most likely little strands of nucleic acids, not DNA yet, which are more sophisticated molecules, but something that could catalyze chemical reactions and that had the blueprint for its own reproduction.

The creation of life was probably not a protracted process using a chemistry that is pretty unlikely but rather is chemistry that occurs rapidly when conditions for the recipe get right. Once the chemistry goes, it goes fairly quickly.

The recipe for life is not that complicated, in that our bodies are made from a limited number of elements. Most of our mass is carbon, oxygen, hydrogen, sulfur, plus some nitrogen and phosphorous. There are a couple dozen other elements that are in there in trace amounts, but to a first approximation, we are just a bag of carbon, oxygen, and hydrogen.

These elements of our bodies came together rather easily because the atmosphere is a bag of carbon, oxygen, and hydrogen as well, even though it is not living.

So the ingredients or recipe for life may be fairly simple, but the question is how does that carbon dioxide in the atmosphere (or methane in an early atmosphere) and water vapor and other sources of hydrogen combine and change from simple, inorganic precursors and become the building blocks of life?

Miller and Urey showed us via their experiment that the chemistry was not some improbable chemistry, but a chemistry that is most likely widely distributed throughout our solar system.

Life is a particular form of chemistry in which the chemicals can lead to their own reproduction. Life isn't somehow different from the rest of the planet. Life is something that emerges on a developing planetary surface as part and parcel of the chemistry of that surface.

Life is really just part of the fabric of a planet like Earth.

The steps were probably as follows:

1. The basic 20 amino acids of life spontaneously formed in the primordial soup as illustrated in the Miller and Urey experiment

2. Groups of these created amino acids and coalesced to form into several types of more complex molecules, including ribosome and RNA

3. RNA developed into DNA

4. DNA underwent Darwinian evolution to achieve the first simple one-cell life

5. The DNA of the one-cell life grew to include an improved ability to sense its need for water and

nutrients in the vicinity of its cell walls and created receptors for the sensing of these ingredients

6. The creation of receptors, sensitive to the cell's needs, began the path to life based on information molecules, thus beginning the escalation of simple cellular life to true life that led to the higher life forms

7. The creation of receptor molecules continued and evolved to have hundreds and then thousands of various receptor molecules sensitive to factors outside the cell, including enzymes and hormone massagers that evolved with higher life forms

I'd like to say a bit more about these listed steps.

The early one-cell life forms have been termed trivial replicators, meaning they were not much more capable of progress than non-life crystals as described earlier. They were capable of processing information only in the passive sense. They relied strictly on basic physics and chemistry of the current environment to support replication; therefore, only a limited set of cellular products was constructible.

But the later cells, which have been termed non-trivial replicators, developed informational molecules and therefore could harness all of the underlying laws of physics and chemistry to achieve a broader agenda for growth and evolution. They process information in an active sense, enabling the possibility of change in response to the current informational state of the system and its surroundings.

The true origin of life did not occur until there was a transition from the trivial replicators to the non-trivial replicators. This occurred when the cellular molecules that had been chemical and structural molecules took on the third role of informational molecules.

We can deduce how this most likely happened by combining two approaches, extrapolating the current properties of modern organisms backwards in time and then deducing the step-by-step evolution of the one-cell trivial replicators to the non-trivial replicators."

In modern organisms, RNA is a biochemical mediator, enabling the translation of DNA to protein. RNA is unique in that it can act as both a genetic polymer and a biochemical catalyst.

This has led to the popular RNA world hypothesis, where all known life is believed to have descended from an ancestral population of organisms that utilized RNA as their sole major biopolymer prior to the advent of DNA and proteins.

This hypothesis had some conceptual problems, but these problems can be solved by amending the hypothesis to recognize that RNA was preceded by an alternative genetic polymer, such as peptide nucleic acid.

Introducing these pre-RNA molecules allows the evolutionary process to be much easier for the sequence of building blocks from lipids, to peptides, and to iron/sulfide complexes because it's easier to synthesize these molecules, and the elements for these products would have been much more abundant on the pre-life Earth.

This sequence approach can construct any possible peptide composed of the amino acids with ribosome acting as the supervisory molecule.

This system sequence would then have undergone further evolution to arrive at the DNA/protein world we observe today.

The key to the varied array of molecules is the enormous number of possible sequences that can be composed from the natural set of the 20 amino acids found in proteins.

And remember that the Miller and Urey experiment showed how the amino acids could spontaneously form.

Following all of this we recognize that the key to life is the ability for the cells to receive instructions telling them what to do.

The signals providing these instructions are mostly chemical in nature.

Simple cells have sensors that detect nutrients, and they help the cells to seek food sources via chemical attractions.

Multi-cellular organisms have developed more complex signals including growth factors, hormones, neurotransmitters, and extracellular matrix components. These substances can exert their effects (attractions) both locally and over long distances.

Neurotransmitters are a class of short-range signaling molecules that travel across the tiny spaces between adjacent neurons or between neurons and muscle cells. Other signaling molecules must move much further to reach their targets. One example is follicle-stimulating hormone, which travels from the mammalian brain to the ovary, where it triggers egg release.

Some cells also respond to mechanical stimuli. For example, sensory cells in the skin respond to the pressure of touch, whereas similar cells in the ear react to the movement of sound waves. In addition, specialized cells in the human vascular system detect changes in blood pressure that the body uses to maintain a consistent cardiac load.

Cells have proteins receptors that bind to the signaling molecules and initiate a physiological response. Different receptors are specific for different molecules. Dopamine receptors bind dopamine, insulin receptors bind insulin, and nerve growth factor receptors bind nerve growth factor, and so on. There are literally hundreds of receptor types in cells, and various cell types have varying populations of receptors.

Receptors can also respond directly to light or pressure, which makes cells sensitive to events in their surroundings and to the atmosphere.

Receptors are generally trans-membrane proteins, which bind to signaling molecules outside the cell and subsequently transmit the signal through a sequence of molecular switches to internal signaling pathways.

There are three major classes of membrane receptors that transform external signals to internal signals: via protein action, via ion channel opening, and by enzyme activation.

Because membrane receptors interact with both extracellular signals and molecules within the cell, they permit signaling molecules to affect cell function without actually having to enter the cell. This is important because most signaling molecules are either too big or too charged to cross a cell's plasma membrane.

Not all receptors exist on the exterior of the cell. Some exist deep inside the cell, and some even exist in the nucleus. These receptors typically bind to molecules that can pass through the plasma membrane, such as gases like nitrous oxide and steroid hormones like estrogen.

Once a receptor protein receives a signal, it undergoes a conformational change, which in turn launches a series of biochemical reactions within the cell. These intracellular signaling pathways typically produce multiple intracellular signals for every receptor that is bound.

At any one time a cell is receiving and responding to numerous signals, and multiple signal transduction pathways are operating in its cytoplasm.

This complexity slowly evolved over long periods of Darwinian time, producing the amazing array of living creatures now on planet Earth.

I guess I have to accept Darwinian evolution, even for the creation of the informational messengers that I had earlier "hoped" were chinks in the theory.

It is very hard to give up on the teachings from boyhood.

Chapter 5
Darwinian Evolution, Follow the DNA

"DNA is like a computer program but far, far more advanced than any software ever created." **Bill Gates**

Perhaps a bit of review is necessary.

I grew up in the Christian world of the Deep South. We were taught stories and lessons from the Bible. God created all life on earth. He created Adam and Eve as the first humans.

But science has proven that man is related to lower life forms.

When fossils of Neanderthals and other species of pre-humans were found, we were given the first clue of this.

But the real proof is in our DNA. We only have to follow the DNA to understand our origins.

When we look at the enormous amount of data on the origins of life and the origins of man, we see that the answers are complex. Did God really create life and man as stated in the Bible? Or did we come into being via evolution? Or did God use evolution as His tool to create life and man?

Let's explore the possibilities.

Humans have approximately 20,000 to 25,000 genes and share 99% of their DNA with the now extinct Neanderthal, and they share 95% to 99% of their DNA with their closest living evolutionary relatives, the chimpanzees.

DNA is Mother Nature's, or if you are a believer, God's primary tool for creating Man and all life forms on planet earth.

After creating the first Hominoid about 6 million years ago, a minor change was made in its DNA to create the next species. This process continued until about 2 million years ago when the genus Homo (our genus) was established.

The process then continued until about 200,000 years ago when the first Homo sapiens came to be, and Cain was born. The process is still continuing today.

The genetic difference between individual humans today averages a minuscule 0.1% (.001).

There is only a 1.2% difference between humans and the chimpanzee.

The DNA difference between gorillas and us is about 1.6%. And I should note that chimpanzees, bonobos, and humans all show this same amount of difference from gorillas.

A difference of 3.1% distinguishes us "African apes" from the Asian great ape, the orangutan.

All of the great apes and humans differ from the rhesus monkeys, for example, by about 7% in their DNA.

The DNA evidence shows that our human creation or evolutionary tree is embedded within the great ape's tree. In the scientific classification we are classified as a great ape.

Don't get upset; it's just the name of a classification.

The fossil evidence supports this DNA evidence, or should I say that this DNA evidence supports the fossil evidence.

Let me summarize.

Due to billions of years of creation or evolution, humans share genes with all living organisms, including plants. The percentage of genes or DNA that organisms share records their similarities. We share more genes with organisms that are more closely related to us.

We have already discussed the very high percentages of DNA that we share with the apes. But we also share high percentages of our DNA with all living creatures. We share 90% with cats, 80% with cows, 75% with mice, 60% with the fruit fly, and 50% with the banana.

Yes, the banana!

Animal and plant life share so much ancient DNA coding because animals and plants had the same ancestors way back and did not diverge until approximately 1.5 billion years ago.

Humans belong to the biological group known as primates, and they are classified with the great apes, one of the major groups of the primate creation or evolutionary tree. Besides similarities in anatomy and behavior, our close biological kinship with other primate species is indicated by DNA evidence. It confirms that our closest living biological relatives are chimpanzees and bonobos, with whom we share many traits.

But we were not created, nor did we evolve directly from any primates living today.

DNA shows that our species and chimpanzees' species diverged from a common ancestor species that lived between 6 and 8 million years ago.

The last common ancestor of monkeys and apes lived about 25 million years ago.

You can get another view of information on this creation or evolution from my book: **Cain's Wife, Lilith's Daughter.**

CAIN'S WIFE
LILITH'S DAUGHTER

BY WALTER PARKS
READ BY ALEX HYDE-WHITE

I wrote this book as part of my struggle between the Biblical teachings of my boyhood and the continuing scientific findings. I needed to know just who Cain's wife was.

You may enjoy what I concluded.

So it is clear that we were either created or evolved from that same seed of life previously described.

The DNA proves this beyond any doubt.

The question is did we evolve by happenstance like the atheists believe, or did God use evolution as His tool to create us?

We know the Biblical story of Adam and Eve is an analogous summary of how we were created; in light of the scientific evidence we know that the Biblical story cannot be literal. We also know that the Bible was significantly rewritten in the late BC and early AD era.

The scientific evidence has continued to grow, especially over recent years, and has caused many who'd blindly accepted the Biblical story in the past to begin to question its literal validity.

Most people have now accepted evolution. It really is fundamental.

Chapter 6
Adam and Eve Chromosomes

"The Y chromosome is passed down identically from father to son, and the DNA from the mitochondria is passed down identically from mother to daughter. We can therefore reveal our paternal and maternal lineage back to Adam and Eve." **Walter Parks**

Okay, so we know that we, beyond a shadow of a doubt, evolved from lower animals. So when did we first become humans?

A very large number of studies and tests have been conducted to date the original human parents of those of us living today.

It turns out that we can trace changes in the Y chromosome of men and the mitochondrial chromosomes of women to determine, at least approximately, when our ancestors first obtained our human genes.

We have named our first two ancestors Y-chromosomal Adam (Y-MRCA) and Mitochondrial Eve, after the Biblical story of creation.

Various study groups have estimated Adam to have been born 188,000, 270,000, 306,000, or 142,000 years ago. A very recent paper reported an older estimate of 338,000 years.

And just as this book is being written there have been two simultaneous reports with one suggesting 180,000 to 200,000 years ago and the other suggesting between 120,000 and 156,000 years.

Analogous to Y-chromosomal Adam, Mitochondrial Eve has been estimated to have lived about 140,000 to 200,000 years ago in Africa. But there have also been several analyses of when she lived.

Theoretically, it is not necessary to believe that Y-chromosomal Adam and Mitochondrial Eve should have lived at the same time. However, very recent findings in 2013 give the possibility that the two individuals could have been contemporaneous.

Analyzing all of the available aging date and comparing it to fossils previously found, I have concluded that Adam and Eve, our first human ancestors were sexually active between 141,000 and 314,000 years ago.

Just to get an answer I averaged these two numbers and concluded that Adam and Eve were sexually active about 228,000 years ago.

The first humans thus most likely lived 228,000 years ago.

Stay tuned for more details in my forthcoming book: **Eden Evolution**.

Eden
Evolution

Walter Parks

Epilogue: Follow Your DNA

You may want to know that you can follow your own DNA back many, many years.

I am going to take a crack at following mine back as far as I can go.

I'll let you know the results as soon as I have them.

I'll also let you know the steps and procedures I found useful so that you too may follow your DNA to create you own ancestral tree.

Stay tuned at info@UnknownTruths.com

About the Author

Hi! Thanks so much for your interest in my books!

My principal interests are true stories of the unusual or of the previously Unknown or unexplained. I occasionally also write fiction.

I was born in Memphis, Tennessee and grew up in Saltillo, Mississippi, a small town near Tupelo, Mississippi. High School life was dominated by watching the rise of our local Elvis. I was editor of my high school paper and had plenty to write about. I guess this was the beginning of my writing career.

After graduating from Mississippi State University as an aerospace engineer I moved to Orlando Florida and worked for Lockheed Martin for 24 years. I advanced from an aerospace engineer to a Vice President of the Company and President of the Tactical Weapons Systems Division.

Educational Activities

I continued my education throughout my career with an MBA degree from Rollins College and with Post Graduate Studies in Astrophysics at UCLA; Laser Physics at the University of Michigan; Computer Science at the University of Miami; Gas Dynamics at MMC and Finance and Accounting at the Wharton School, University of Pennsylvania.

While at Mississippi State University, I was on the President's honor list and in the honor societies of Tau Beta Pi, Sigma Gamma Tau, and Blue Key.

I received a scholarship from Delta Air Lines based on my academics and performance.

I was in ROTC and the Arnold Air Society where I participated and toured as a member of the precision drill team. I also attended the summer survival training at Hamilton Air Force Base in California.

I was selected for Who's Who among Students in American Universities and Colleges.

I was a speaker for several technical organizations, including the American Institute of Aeronautics and Astronautics.

After Retirement from Lockheed Martin Aerospace Company

After retirement from Lockheed, I formed Parks-Jaggers Aerospace Company and sold it four years later.

After selling my aerospace company I formed Quest Studios, Quest Entertainment, and Rosebud Entertainment to make films at Universal Studios. I produced ten films, directed seven films, and wrote five film scripts produced at Universal Studios.

I won the National Association of Theater Owners Show South Producer of Tomorrow Award.

I then formed UnknownTruths Publishing Company to publish true stories of the unusual or of the previously Unknown or unexplained. These include books about past events so unbelievable that most people have relegated them to myths.

I have published 30 books with 28 in eBook format, 21 in Paperback format, and 23 as Audio Books. The ones most related to **First Life on Earth, Scientific Adam & Eve** are:

Life and the Universe

Cain's Wife Lilith's Daughter

Hormones Working for You

I have an additional 12 books in development.

My Key Website:
UnknownTruths.com
My Amazon Author Page:
http://www.amazon.com/-/e/B00EE9HWA4

About
UnKnownTruths
Publishing Company

UnKnownTruths Publishing Company was formed to publish true stories of the unusual or of the previously Unknown or unexplained. These stories typically provide radically different views from those that have shaped the understandings of our natural world, our religions, our science, our history, and even the foundations of our civilizations.

The Company's stories also include stories of the very important anti-aging, life-extending medical breakthroughs, stem cell therapies, genetic therapies, cloning, and other emerging findings that promise to change the very meaning of life.

The Company also publishes stories from the past that are so unbelievable that they are generally considered to be myths. The published stories provide the evidence for the truth.

The Company has published 30 books with 28 in eBook format, 20 in Paperback format, and 23 as Audio Books. The Company currently has an additional 12 books in development.

UnknownTruths.com

Please Write a Review

Go to: http://www.amazon.com/-/e/B004S7JLBA

Then scroll down to see where to write the review.

Please Visit

UnknownTruths.com

To Opt-In and get FREE eBooks and other Freebies.